A Guide to Rearing Swans

The ultimote handbook to get the best out of your Swans

Chapter one

Introduction

Welcome to "Among the Swans: A Guide to Rearing Swans"! Prepare to enter the enchanting world of these majestic waterfowl as we embark on a journey to uncover the wonders of raising swans.

Imagine gazing upon a serene lake, surrounded by the graceful presence of swans gliding effortlessly across the tranquil waters. Swans, with their elegant plumage, regal appearance, and soothing demeanour, have captivated humans for centuries. And now, you too can embark on an extraordinary adventure of rearing and caring for these magnificent creatures.

In this comprehensive guide, we will delve into the captivating beauty of swans, exploring their different types, their symbolism in art and literature, and their remarkable physical attributes. We will then delve into the practical aspects of rearing swans, covering the essentials such as selecting the perfect location, providing proper food and water, ensuring impeccable sanitation, and safeguarding them from predators.

But this guide offers more than just practical advice. We will also unravel the intriguing behaviours of swans, discovering how they communicate, socialize, and engage in complex mating rituals. By understanding their behaviour, you'll establish a deeper connection with your swans, unlocking the secrets to fostering a harmonious relationship.

So whether you are a nature enthusiast, a passionate bird lover, or simply someone seeking a unique and rewarding experience, join us as we embark on this remarkable journey "Among the Swans." With our guide in hand, you will be equipped with the knowledge and skills to create a haven for these regal birds and witness firsthand the absolute splendour of rearing swans. Let's dive in!

1.1 Importance of Swans as Graceful Waterfowl

Swans are the epitome of grace and beauty and are some of the most recognizable and admired waterfowl in the world. These iconic birds are widely celebrated for their stunning elegance, majestic stature, and strikingly beautiful features. Apart from their undeniable charm, swans also hold significant ecological, cultural, and historical importance. In this section, we will explore the different ways in which swans are essential waterfowl, and the role they play in our society.

Ecological Importance

Swans are an integral part of our freshwater and wetland ecosystems. As herbivores, they feed on aquatic plants, phytoplankton, and invertebrates, and play a crucial role in maintaining the wetland and aquatic ecosystems' health. Through their feeding habits, they help control the growth of algae and other aquatic plants, which can cause imbalances in the ecosystem. The swans also serve as mobile nutrient carriers, distributing the nutrients they consume throughout their habitat. They also aid in the propagation of aquatic plants by helping to disperse seeds and tubers throughout the habitat, ensuring the continued growth and diversity of the plant community.

Swans also play a vital ecological role in the food chain, with both adults and cygnets providing prey for larger predators such as foxes, raccoons, and coyotes. They also attract a variety of bird species to their habitat, providing birdwatchers and other wildlife enthusiasts with an opportunity to observe a diverse range of bird species.

Cultural and Historical Significance

Swans have long been revered and celebrated in human cultures worldwide, with depictions of swans dating back to prehistoric times. Swans were highly prized and coveted in medieval Europe and feature prominently in folklore, art, and literature. According to Greek and Roman mythology, swans were associated with the gods Apollo and Venus.

In England, swans were declared royal birds in the 12th century, and the tradition of swan-upping (counting and marking of swans) continues to this day on the River Thames. In China, swans hold a special place of reverence and are associated with wisdom, grace, and beauty. They are depicted in traditional Chinese painting, embroidery, and literature.

Touristic Attraction

Swans are also a significant tourist attraction in many parts of the world. Swans are commonly seen in parks and public spaces and are used to enhance the aesthetic appeal of these areas. In some countries, swans are even used as a symbol of the city or region where they reside, and their presence attracts visitors and tourists looking to experience the beauty and tranquillity of these elegant waterfowl.

1.2 Benefits of Rearing Swans

Rearing swans can be an excellent way to connect with nature and add aesthetic value to your property. While swans can be challenging to care for, they also offer several benefits that can make it worth the effort. In this section, we'll explore some of the benefits of rearing swans.

Aesthetics

Swans are elegant and graceful birds that can add beauty and sophistication to your property. Swans are commonly found on large bodies of water such as lakes, ponds, and rivers, and their presence can enhance the natural beauty of these areas. Swans' white plumage and long necks are visually appealing and are often depicted in paintings, sculptures, and other forms of art. If you own a property with a lake or pond, rearing swans can be an excellent way to enhance its aesthetic appeal.

Attractiveness to Visitors and Wildlife

Swans can be an attractive sight for visitors to your home or property. People are naturally drawn to the beauty and grace of swans, and having them nearby can provide a sense of serenity and tranquility. Moreover, swans can attract other wildlife to the area, making your property a hub of natural activity. Other birds, fish, and small animals may be drawn to the swans, which can create a diverse and thriving ecosystem.

Emotional and Therapeutic Benefits

Swans can have a calming and therapeutic effect on people. Studies suggest that watching birds can help lower blood pressure, reduce anxiety and stress, and provide a sense of peace and relaxation. Swans, in particular, are known for their serene and tranquil presence and can offer emotional benefits that are hard to find elsewhere. Whether you're enjoying the swans' company on a summer afternoon or watching them glide gracefully across the water, you're sure to find peace and relaxation in their quiet beauty.

Educational Opportunities

Rearing swans can provide unique educational opportunities for both children and adults alike. Children can learn about the natural world and the role swans play in their environment. Adults can also learn about swans' behaviours, ecology, and habitat, which can be an enriching and educational experience. Furthermore, rearing swans can provide opportunities for community involvement such as organizing birdwatching events or leading educational programs for local schools.

Conservation and Preservation

Swan rearing can also contribute to conservation and preservation efforts. Swans are an essential part of the natural ecosystem, and by rearing them, you're contributing to their preservation and protection. You can also help raise awareness for conservation efforts and educate others on the importance of swan preservation.

Chapter two

The Beauty of Swans

2.1 History and Symbolism of Swans

Swans have a rich history and deep symbolism that spans across cultures and centuries. Revered for their beauty and grace, swans have captivated the human imagination and have been an enduring symbol of various qualities and concepts. In this section, we will explore the fascinating history and symbolism of swans.

Ancient Mythology and Legends

Swans have a significant presence in ancient Greek and Roman mythology. In Greek mythology, swans were associated with the gods Apollo and Aphrodite, among others. According to a popular myth, Apollo was said to transform into a swan to pursue his love interests. This association with the gods elevated the swan to a status of divine beauty and elegance.

In Celtic mythology, swans were often seen as mystical creatures and were believed to possess the ability to travel between the earthly world and the afterlife. They were associated with the realm of the Otherworld and were considered symbols of purity and spiritual transformation.

Royal Birds and Aristocratic Symbolism

In medieval Europe, swans held a special status and were closely associated with royalty and nobility. In England, swans were declared royal birds in the 12th century by law. The ownership and care of swans became a privilege of the wealthy and the crown. This association with the aristocracy further elevated the swan's image, making it a symbol of elegance, sophistication, and refinement.

Symbol of Love and Romance

Swans are often regarded as symbols of love and romance. In many cultures and folklore, swans are associated with fidelity, devotion, and lifelong partnership. Swans mate for life and are known for their faithfulness to their partners. This enduring bond has made them symbols of eternal love and romantic relationships.

Grace and Tranquility

Swans are renowned for their graceful movements and serene presence on the water. Their elegant necks, ethereal white feathers, and majestic stature have made them natural symbols of grace, serenity, and tranquillity. Artists and poets throughout history have sought inspiration from swans to convey a sense of harmony and inner peace.

Symbol of Rebirth and Transformation

Swans are often associated with the concept of rebirth and transformation. Their life cycle, from egg to cygnet to fully grown adult, mirrors the process of growth and change. This symbolism is particularly prevalent in Chinese culture, where swans are seen as creatures of wisdom and beauty, and they represent transformation and the pursuit of enlightenment.

Spiritual and Mystical Symbolism

In various spiritual traditions, swans are seen as symbols of higher consciousness and spiritual awakening. Their ability to glide effortlessly on water and soar through the skies is seen as a metaphor for the soul's journey and transcendence. Swans are associated with themes of purity, enlightenment, and inner wisdom, often depicted in spiritual and esoteric artwork.

2.2 Physical Characteristics and Features

Swans are large and majestic birds known for their elegant appearance and graceful movements. With their long necks, beautiful plumage, and distinctive features, swans are a captivating sight in the natural world. In this section, we will explore the physical characteristics and features that make swans unique.

1. Size and Stature

Swans are among the largest flying birds, with an average length ranging from 1.2 to 1.5 meters (4 to 5 feet). Their wingspan can span up to 2.4 meters (almost 8 feet), allowing them to soar through the air with remarkable ease. Adult swans can weigh between 7 and 15 kilograms (15 to 33 pounds), with males often being slightly larger than females.

2. Plumage and Coloration

One of the distinguishing features of swans is their beautiful plumage. Most swan species have predominantly white feathers, which give them an ethereal and pristine appearance. The white colouration is caused by pigments in their feathers, and it acts as camouflage in their natural habitat of wetlands and bodies of water. The contrasting black eyes of swans add to their striking presence.

In some species, such as the black swan, the feathers can have a black or gray shade mixed with white. During the breeding season, swans may develop various colorations and patterns on their feathers as a part of their courtship or territorial displays.

3. Long and Graceful Neck

Another prominent feature of swans is their long and flexible necks. Swans possess a varying number of neck vertebrae, allowing them to move their necks in a graceful and sinuous manner. Their long necks are not only visually striking but also provide practical benefits. Swans use their necks to reach underwater vegetation and to preen their plumage, ensuring their feathers remain clean and waterproof.

4. Beak and Facial Characteristics

The beak of a swan is large, strong, and generally black in colour. The shape of the beak varies slightly between species but is usually long and rounded. Swans use their beaks to forage for food by probing in the mud or water for vegetation, insects, and small aquatic creatures.

A notable facial feature of swans is the fleshy knob, called a "blackberry," located at the base of their beaks. This knob is more prominent in males and can develop during mating season. The purpose of this knob is not entirely understood, but it is believed to play a role in courtship and territorial displays.

5. Webbed Feet

Swans have strong and webbed feet, adapted for swimming and navigating through aquatic environments. These specialized feet allow them to propel themselves through the water with ease. The webbing between their toes enables efficient swimming and provides stability while walking on muddy or uneven surfaces.

6. Flight and Wings

Despite their size, swans are proficient flyers. Their large wingspan and powerful muscles enable them to take flight and navigate the skies gracefully. While they may appear slow and heavy on land, they can swiftly and gracefully glide through the air. Swans use their wings to generate lift and achieve flight, often seen in V-formation flights during migration.

2.3 Elegance and Gracefulness

Swans are renowned for their elegance and gracefulness, captivating observers with their serene presence and effortless movements. From their majestic appearance to their synchronized swimming, swans possess a natural beauty that has been celebrated and admired throughout history. In this section, we will explore the elegance and gracefulness that defines swans.

Majestic Appearance

Swans are known for their regal and majestic appearance. With their long necks held high, rich plumage, and striking postures, swans exude an air of dignity and poise. They embody the grace and beauty that has captivated artists, poets, and writers for centuries.

Serene Presence

Swans have a serene presence that can create a sense of tranquillity in their surroundings. Whether they are gracefully gliding across a still lake or quietly preening their feathers on the riverbank, swans exude a calming and peaceful presence. Their serene demeanour and unhurried movements invite observers to pause and appreciate the beauty of the moment.

Synchronized Swimming

One of the most captivating displays of elegance and gracefulness by swans is their synchronized swimming. When swimming in groups or pairs, swans move in perfect harmony, creating an exquisite display of coordination and fluidity. With synchronized strokes and carefully coordinated movements, they effortlessly travel through the water, creating ripples and leaving a lasting impression on those who witness this remarkable sight.

Graceful Neck Movements

Swans are known for their gracefully swan-like neck movements, a characteristic that adds to their overall elegance. They can curve and contort their necks into beautiful and sinuous shapes, enhancing their allure and capturing the attention of onlookers. Whether stretching their necks to reach underwater vegetation or gracefully preening their feathers, swans' neck movements are a visual representation of their elegance and gracefulness.

Delicate Flight

Despite their size, swans exhibit a delicate and graceful flight. With their large wingspan and strong wing muscles, they take to the skies with remarkable ease and finesse. Their flight is characterized by smooth and gliding movements, as they soar through the air with elegance and precision. Whether seen in solitary flights or in V-formation during migration, swans inspire awe and admiration with their graceful aerial displays.

Symbolic Representation of Elegance

Swans have become symbols of elegance and gracefulness in various cultures and traditions. Their effortless movements and dignified presence have made them metaphors for beauty, refinement, and sophistication. In literature, art, and folklore, swans often represent ideals of elegance and grace, inspiring human beings to strive for these qualities in their own lives.

Chapter three

Types of Swans

3.1 Mute Swans

Mute swans (Cygnus olor) are exquisite waterbirds known for their elegance and striking appearance. With their beautiful white plumage, long necks, and distinctive orange beaks, they have captured the attention and fascination of humans for centuries. In this section, we will delve into the world of mute swans, exploring their physical descriptions and unique characteristics that set them apart from other avian species.

Description

Mute swans are large birds, measuring approximately 1.5 to 1.8 meters (5 to 6 feet) in length and possessing a wingspan of about 2.4 to 2.6 meters (8 to 8.5 feet). They are well-known for their graceful and curved necks, which add to their overall majestic appearance. Their necks are both striking and distinct, bending in elegant arcs that captivate onlookers.

One of the most noticeable characteristics of mute swans is their pure white plumage. This vibrant colour extends down to their wings, encompassing their entire body. The consistent colouration acts as a form of camouflage in the habitats they frequent, enabling them to blend in with the surrounding water bodies, such as lakes, rivers, ponds, and estuaries.

Beak and Knob

The beaks of mute swans are a distinguishing feature. They are vibrant orange in colour and possess a graceful curvature. The beak's shape enables the swans to feed on aquatic vegetation and invertebrates, calmly foraging for their sustenance. At the base of their beaks, mute swans have a black knob, which becomes more prominent as they reach maturity. This knob adds an additional touch of uniqueness to their appearance.

Monogamous Nature

Mute swans are monogamous birds, known for forming lifelong pair bonds with their chosen mates. These pairs engage in beautiful courtship displays, which involve various elegant movements and synchronized swimming. During these displays, the pairs engage in mutual head and neck movements, often accompanied by flapping their wings. These courtship rituals not only strengthen their pair bond but also serve as territorial and mating rituals.

Nesting and Parental Care

When it comes to nesting, mute swans build large and elaborate nests made primarily of vegetation, typically near the water's edge. The female lays 4 to 7 eggs, which are incubated by both parents for approximately 35 to 41 days. The parents take turns incubating the eggs and caring for the nest, sharing the responsibilities equally. Once the cygnets (baby swans) hatch, they are cared for by both parents until they are ready to venture out on their own.

Territorial Behavior

Mute swans are highly territorial birds, particularly during the nesting season. They fiercely defend their nesting territory against intruders, using their wings and bills to ward off potential threats. Their territorial behaviour showcases their strength and determination, adding to their reputation as powerful and protective creatures.

Graceful Movements

One of the most enchanting aspects of mute swans is their graceful movements. They glide effortlessly across the water's surface, their long necks held high, exuding an air of tranquility and serenity. These graceful movements are admired by onlookers, who find solace in their gentle presence.

Adaptability and Resilience

Mute swans exhibit adaptability and resilience in various environments. They are highly adaptable to different types of water bodies, from freshwater lakes to saltwater estuaries. Their feeding habits also demonstrate their adaptability, as they can consume a wide range of aquatic vegetation and invertebrates found in their habitats.

Conservation Concerns

While mute swans are widely distributed across Europe, Asia, and North America, they face conservation concerns in certain regions. Habitat loss, pollution, and the degradation of wetlands have impacted their populations in some areas. Additionally, the introduction of invasive species has disrupted the delicate balance of ecosystems where mute swans reside. Conservation efforts are crucial to ensuring the long-term survival and well-being of these majestic creatures.

3.2 Black Swans

Black swans (Cygnus atratus) are enigmatic and captivating waterbirds that possess a rare and striking physical appearance. With their glossy black plumage, vibrant red beaks, and elegant necks, they command attention and intrigue. In this section, we will explore the world of black swans, delving into their descriptions, characteristics, and the unique allure they bring to the avian realm.

Description

Black swans are large birds, measuring approximately 1.2 to 1.4 meters (4 to 4.6 feet) in length, with a wingspan ranging from 1.6 to 2 meters (5.2 to 6.6 feet). They are slightly smaller than their white swan counterparts but possess a distinctive and eye-catching appearance. A notable feature of black swans is their rich and lustrous black plumage, which serves as a stark contrast against their vibrant red beaks and white feather lines that run along the edges of their wings.

Plumage and Coloration

The plumage of black swans is an embodiment of elegance and mystery. The feathers appear velvety and have a glossy sheen that enhances their allure. The deep black colour of their plumage allows them to blend seamlessly into their preferred habitats, which include lakes, rivers, and wetlands. Unlike other species of swans, black swans do not possess pure white feathers, making their black plumage stand out even more vibrantly.

Red Beak and White Feather Lines

One of the most striking features of black swans is their vibrant red beaks. This vibrant hue adds a splash of colour to their dark plumage, creating a visually arresting contrast. The beak itself is long and slender, perfectly complementing the elegance of their overall appearance.

A distinguishing characteristic of black swans is the thin white feather lines that run along the edges of their wings. These stark white lines provide an additional layer of visual interest, breaking the monotony of their black plumage and offering a unique touch of beauty.

Territorial Nature

Black swans are known for their territorial behaviour, especially during the breeding season. They establish and fiercely defend their breeding territories, working diligently to protect their nesting sites and ensure the safety of their young. When intruders encroach upon their territory, black swans can display aggression, flapping their wings and vocalizing loudly to ward off any potential threats.

Adaptability and Range

Black swans are native to Australia and were once considered a rare and iconic symbol of the continent. However, they have since been introduced to various countries around the world, where they have established themselves successfully. They display remarkable adaptability, thriving in a range of habitats, from coastal regions and estuaries to freshwater lakes and ponds.

Sociability and Pair Bonding

Black swans are highly sociable birds, often seen congregating in small groups or pairs. They form strong pair bonds, engaging in displays of mutual grooming and synchronized swimming. Mating pairs can remain together for life, nurturing a deep commitment and sense of partnership.

Conservation Status

Black swans have a stable population, and their conservation status is considered of "Least Concern" according to the International Union for Conservation of Nature (IUCN). Their adaptability and successful establishment in various habitats contribute to their overall resilience. However, it is crucial to remain vigilant in protecting their habitats and ensuring the continued conservation of their natural environments.

Native Habitat and Distribution

Black swans (Cygnus atratus) are renowned for their striking appearance and captivating beauty. They are native to Australia and have established a strong presence in various parts of the continent. In this section, we will delve into the native habitat and distribution of black swans, exploring the environments they call home and the regions where they can be found.

Native Habitat

Black swans are primarily found in Australia, where they are deeply entwined in the country's cultural identity. They have adapted to a diverse range of habitats, showcasing their resilience and ability to thrive in various environments. Here are some of the key native habitats where black swans can be commonly observed:

1. Wetlands

Black swans are frequently found in wetland areas, including both freshwater and saltwater wetlands. These habitats offer an abundant supply of aquatic vegetation and invertebrates, which form a substantial part of their diet. Wetlands serve as ideal locations for nesting, as they provide ample cover and resources for the swans to reproduce and raise their offspring.

2. Lakes and Rivers

Black swans are commonly seen in lakes and rivers across Australia, where they glide gracefully across the calm waters. These habitats provide suitable conditions for the swans to feed on aquatic plants and forage for small aquatic animals. Lakes and rivers also offer nesting opportunities, as the swans can build their nests along the water's edge, ensuring a safe and accessible environment for their young.

3. Coastal Regions

Coastal regions play host to black swans, attracting them with their mix of saltwater and estuarine habitats. Here, the swans can find a diverse range of food sources, including seagrasses, algae, and small crustaceans. Coastal areas also offer ample nesting sites, with sand dunes and coastal vegetation providing suitable locations for the construction of their nests.

Distribution in Australia

Black swans have a widespread distribution across Australia, and their presence can be observed throughout the continent. Some key regions where black swans can be commonly found include:

Western Australia

The black swan holds special significance in Western Australia, where it is the state's emblem. It is widespread along the coast and inland water bodies, such as the Swan

River in Perth. These swans can often be seen gliding gracefully on the water or foraging in wetland areas.

Victoria

The black swan is also a notable resident of Victoria, particularly in regions such as the Gippsland Lakes and the Melbourne metropolitan area. Wetlands, lakes, and coastal areas in Victoria provide suitable habitats for black swans to thrive and breed.

South Australia

In South Australia, black swans can be found in various coastal and inland locations, including the Coorong, Lower Lakes, and Murray River regions. These areas offer the necessary resources and nesting sites for black swans to establish their presence.

New South Wales and Queensland

Black swans can be observed in New South Wales and Queensland, where they inhabit a range of habitats, from coastal estuaries and lakes to inland rivers and wetlands. The Tweed River region and the Moreton Bay area are particularly notable for their black swan populations.

Northern Territory

In the Northern Territory, black swans are found in wetland areas, including the Kakadu National Park and other freshwater habitats. These regions provide the swans with the resources they need to thrive and successfully reproduce.

Conservation Efforts

Black swans maintain a stable population throughout their native range in Australia, and their conservation status is currently listed as "Least Concern" by the International Union for Conservation of Nature (IUCN). However, it remains essential to protect their habitats and the ecosystems they rely on, ensuring the continued existence of these magnificent birds for future generations to appreciate and admire.

Unique Features and Behavior

Black swans (Cygnus atratus) are fascinating creatures with a range of unique features and behaviors that set them apart from other waterbird species. From their striking appearance to their intriguing social dynamics, in this section, we will explore the distinctive characteristics and behaviors that make black swans truly captivating.

Striking Plumage

Black swans are known for their stunning plumage, which consists of deep black feathers that give them their name. Unlike their white swan counterparts, their feathers lack any hint of white, creating a visually striking appearance. The glossy texture of their plumage adds to their allure, with the feathers shimmering under sunlight, enhancing their visual impact.

Vibrant Red Beak

One of the most striking features of black swans is their vibrant red beak. The intense red coloration serves as a stark contrast against their black plumage, making it a prominent feature that draws attention. The beak itself is long, slender, and elegant, perfectly complementing the overall grace of the swan's appearance.

White Feather Lines

Adding to the visual appeal of black swans are the thin white feather lines that run along the edges of their wings. These lines provide a beautiful contrast against their black plumage, breaking up the monotony and adding a touch of elegance. The white feather lines stand out prominently, creating a visually striking pattern as the swan glides through the water or spreads its wings in display.

Impressive Size and Elegance

Black swans are large birds, measuring approximately 1.2 to 1.4 meters (4 to 4.6 feet) in length, with a wingspan ranging from 1.6 to 2 meters (5.2 to 6.6 feet). Despite their size, they possess a graceful and elegant demeanour, moving with a sense of poise both on land and in water. Their long neck, sleek body, and smooth movements contribute to their overall aura of grace and elegance.

Territorial Behavior

Black swans are known for their territorial nature, especially during the breeding season. They are highly protective of their nesting sites and will vehemently defend their territory against intruders. To ward off potential threats, black swans engage in aggressive displays, flapping their wings, vocalizing loudly, and even forcefully striking at intruders with their beaks. This territorial behavior showcases their commitment to safeguarding their young and ensuring the survival of their offspring.

Sociability and Pair Bonding

Despite their territorial tendencies, black swans are highly sociable birds, often seen congregating in small groups or pairs. They form strong pair bonds, and once a pair is formed, it tends to remain together for life. The pair engages in various bonding rituals, including mutual grooming and synchronized swimming, strengthening their partnership and reinforcing their commitment.

Cooperative Nesting

Black swans engage in cooperative nesting, with both the male and female participating in the construction of the nest. The nest is typically built on the edges of wetlands or along the water's edge, constructed from a combination of reeds, grasses, and other available vegetation. Both parents take turns incubating the eggs and caring for the hatchlings, sharing the responsibilities of parenting.

Vocalizations

Black swans are not known for their melodic vocalizations like some other bird species, but they do communicate through various sounds. During courtship and territorial displays, they emit a range of vocalizations, including honks, trumpets, and hisses. These vocalizations serve as a means of communication, conveying messages to other swans and warning potential intruders to stay away.

Adaptability

Black swans display a remarkable ability to adapt to various habitats. They can be found in a diverse range of environments, including lakes, rivers, wetlands, estuaries, and even coastal areas. This adaptability allows them to thrive in different conditions and utilize a variety of food sources, ensuring their survival across a wide range of habitats.

3.3 Trumpeter Swans

Trumpeter swans (Cygnus buccinator) are among the largest waterfowl species in the world. Named for their distinctive, trumpet-like call, these swans are revered for their elegance, strength, and captivating beauty. This section will delve into the fascinating aspects of their physical attributes and notable characteristics.

Description and Physical Attributes

Trumpeter swans have several features that distinguish them from other swan species. From physical size to colouration and communicative signals, below are the significant aspects that define these majestic birds:

Size

Adult Trumpeter swans are impressive in size, often reaching lengths of about 138-158 cm (54-62 inches). The wingspan of these magnificent creatures can extend from 185 to 250 cm (73-98 inches), which makes for an awe-inspiring sight when they spread their wings in flight. Their large size, coupled with their pure white plumage, grants them a regal presence and earns them a significant spot in aquatic habitats.

Weight

Trumpeter swans are heavyweight champions among North American waterfowl, with males typically weighing between 11.8 kg and 12.7 kg (26-28 lbs) and females measuring slightly smaller, between 9.6 kg and 10.3 kg (21-23 lbs). Their notable weight doesn't hamper their ability to fly, yet it does make their take-offs appear laborious, featuring a long and powerful run-up along the surface of the water.

Plumage

Trumpeter swans are a vision of pure elegance with their all-white plumage. The pristine white continues throughout the year, although it may become slightly stained in some habitats. Unlike many other waterfowl, these swans do not exhibit any significant sexual dimorphism, meaning males and females have similar external appearances.

Beak and Feet

An easily recognizable attribute of the Trumpeter swan is its black bill, contrasting sharply with its white feathered body. The bill, large and wedge-shaped, has a notably straight top line and a thin red border around the mouth, adding a subtle dash of color. Their feet are also black, which are webbed for efficient swimming.

Eyes

Their eyes are another interesting feature, exuding an intense focus. They are dark and small, giving them a keen sense of their surroundings, crucial for their survival in the wild.

Neck

Trumpeter swans have a long, muscular neck, which aids in feeding and preening. The neck's length also plays a crucial role in thermoregulation, helping the birds to lose or retain body heat as needed.

Voice: The Trumpeting Call

True to their name, Trumpeter swans have a loud, trumpet-like call. This call, described as a low and resonant "ko-hoh", can carry over several kilometers, playing a vital role in communication, establishing territories, and maintaining pair bonds. Their vocal performance is one of their most distinctive features, creating a captivating symphony in their habitats.

Conservation and Preservation

Historically, Trumpeter swans were hunted extensively for their feathers, leading to a severe decline in their population. However, conservation efforts over the past several decades have been successful in gradually increasing their numbers. The species is now protected in many regions, ensuring the survival and longevity of these remarkable birds.

Natural Habitat and Range

Trumpeter swans (Cygnus buccinator), widely known for their elegant beauty and resonant call, have claimed a particular niche in the biodiverse world of North American waterfowl. Their natural habitats and range reflect their adaptability and illustrious presence across various ecosystems. In this section, we will explore the remarkable habitats in which these birds establish their territories and the expansive range they're known to cover.

Natural Habitat

Trumpeter swans comfortably inhabit a diverse set of environments based on the availability of food, nesting sites, and open waters. The following ecosystems are where these magnificent birds can typically be found.

Wetlands

Wetlands form the primary habitat for Trumpeter swans. These include marshes, ponds, and slow-moving rivers, offering generous sources of aquatic vegetation, invertebrates, and mollusks that constitute a significant part of their diet. Besides providing food, wetlands also equip the swans with ideal locations for nesting- broad, undisturbed stretches with abundant vegetation.

Lakes

Trumpeter swans also find solace in the still, expansive vistas of lakes. Lake areas serve up a vibrant menu of rooted aquatic plants and small aquatic animals – an inviting feast for the swans. The well-vegetated lake shores become comfortable spots for them to build nests and raise their cygnets.

Estuaries

Estuaries, or the transitional areas where rivers meet the sea, offer another welcoming environment for the swans. With a mix of fresh and saltwater biomes, estuaries provide a variety of food sources including seagrasses, tubers, and algae catering to the nourishment needs of these birds.

Distribution and Range

Historically, Trumpeter swans inhabited a vast range across North America, from the Arctic coast down to the northern parts of the United States. Over time, due to hunting pressures and habitat loss, their range reduced drastically. However, through successful conservation efforts, their population has started recovering, and they have re-inhabited portions of their historic range. Key geographical areas where Trumpeter swans can be seen today include:

North-Western United States

The Greater Yellowstone Region, specifically the states of Idaho, Montana, and Wyoming, hosts one of the principal populations of Trumpeter swans. The trumpeters here, named the 'Rocky Mountain Population,' are drawn to the area's abundant rivers, lakes, and wetlands.

Northern United States and Central Canada

The 'Interior Population' of Trumpeter swans inhabits the locales such as Iowa, Minnesota, Wisconsin, and Ontario. This population, which almost went extinct in the early 1930s, has seen a resurgence through reintroduction programs and continual conservation efforts.

Pacific Coast

In the Pacific Coastal region, particularly in British Columbia and Alaska (the 'Pacific Coast Population'), Trumpeter swans enjoy abundant wetlands and estuaries. These swans undertake substantial migrations from their breeding grounds in Alaska to their wintering habitats in the coastal areas of southern British Columbia.

Northern Great Plains

Recent conservation efforts have lead to the reintroduction of Trumpeter swans to their historic range in the states of North Dakota, South Dakota, Nebraska, and Kansas. Although still growing, this population is a testament to the resilience and adaptability of these birds.

Conservation Efforts

Although the Trumpeter swan's conservation status is currently listed as "Least Concern" by the International Union for Conservation of Nature (IUCN), it remains vitally important to ensure their habitats and ecosystems they depend upon continue to thrive. Protection of wetlands, sustainable land-use practices, and education about the vital role these birds play in biodiversity, contribute to the conservation of these majestic creatures.

Behaviour and Conservation Efforts

Trumpeter swans (Cygnus buccinator) are not only impressive in size and elegance but are also intriguing in terms of their behavioural traits and the conservation efforts that aid their survival. This section will explore the aspects of their behaviour and the significant conservation efforts aimed to ensure their long-term survival.

Behaviour of Trumpeter swans

Trumpeter swans exhibit a range of behaviours that have evolved to promote survival and reproductive success. These behaviours include their mating patterns, territoriality, migratory habits, and communication methods.

Mating Patterns

Trumpeter swans engage in monogamous relationships that last for life. Pairs form bonds through a courtship ritual that involves synchronized swimming, head-bobbing, and the presentation of aquatic vegetation. Their bonds are highly social and cooperative, with both parents participating in nest building, incubation, and care for the young.

Territorial Traits

Trumpeter swans are fiercely territorial during the breeding season. They protect their nesting sites from potential threats and intruders, striking them with their powerful wings or charging at them. Outside of the breeding season, they are a bit more sociable and can gather in small groups or flocks in wintering sites.

Migration Habits

Depending on their geographical range, some trumpeter swan populations are migratory. For instance, swans breeding in Alaska and other northern parts of their range undertake long-distance migrations to wintering areas on the Pacific Coast. They utilize traditional flight paths, often following the same route year after year.

Communication

Communication within the swan community is far from silent. Trumpeter swans communicate using a range of vocal and non-vocal signals. The trumpeting call, from which they earn their name, is regarded as a primary form of communication. This low and resonant call can be heard over several kilometers.

Conservation Efforts

Trumpeter swans were the subject of extensive hunting during the 19th and early 20th centuries, where they were valued for their skin, feathers, meat, and eggs. This led to a drastic decline in their population, bringing them almost to the brink of extinction in much of their range. But the tale of Trumpeter swans is one of successful conservation, recovery, and protection.

Legal Protection Measures

In the early 20th century, legislative measures were implemented to provide legal protection to Trumpeter swans. Hunting was banned, providing these birds with a critical respite to recover their population numbers.

Habitat Preservation

Efforts have been made to preserve and restore the natural habitats of Trumpeter swans, including wetlands and bodies of water. This restoration provides them with the necessary space to replicate and access to adequate food supply.

Reintroduction Programs

Reintroduction programs have been successful in re-establishing Trumpeter swans to areas from which they had been extirpated. Captive-bred swans have been released into suitable habitats within their historic range, contributing to the population's growth and expansion.

Public Education and Awareness

Public education and awareness campaigns have been instrumental in rallying support for Trumpeter swan conservation. These efforts have focused on shifting public attitudes towards wildlife conservation and promoting responsible behaviours that don't disturb the birds, their nests, or their habitats.

Chapter four

How to Rear Swans

4.1 Choosing a Suitable Location

Rearing swans, such as the magnificent Trumpeter swans (Cygnus buccinator), can be a rewarding venture, providing a unique opportunity to appreciate these birds' beauty and behaviour from up close. However, it requires considerable planning, care, and an appropriately chosen rearing site that caters to their specific needs. This section will delve into some of the critical factors to consider when selecting a suitable site to rear swans.

1. Suitable Water Bodies

Swans, being waterfowl, need access to a body of water in their environment. The water body should be large enough for the swans to swim, bathe, and carry out their natural behaviours. An ideal scenario would be a pond or a lake, which should be at least half an acre in size. The water body should have a mix of deep and shallow areas, allowing the swans to feed and swim comfortably.

2. Food Supply

When considering a swan rearing site, you need to ensure an abundant and consistent food supply. Swans primarily feed on aquatic plants, small fish, insects, and grains. Therefore, the water body and its surroundings should be fertile and rich in these natural food sources. Alternatively, supplemental feeding might be required, which could include poultry layer pellets, lettuce, or corn.

3. Nesting Sites

Swans require safe and quiet places for nest building. The rearing site should offer plenty of spaces that can provide such an environment. Islands or slightly raised mounds within or near the water body can serve as ideal nesting sites. These locations should be away from regular human activity or potential predators.

4. Shelter from Extreme Weather Conditions

Weather conditions play an essential role in determining the suitability of a swan-rearing site. Swans require sheltered spots to protect themselves from strong winds, frosts, or extreme heat. The site should ideally provide natural shelter such as overhanging trees, shrubs, or man-made structures like shelters or trees for roosting.

5. Safety from Predators

The rearing site must be safe from potential predators that could pose a threat to the swans or their eggs. In some geographical regions, this could include large birds of prey, foxes, raccoons, or even dogs. Measures such as fencing, predator deterrents, or a guardian dog might be considered to ensure safety.

6. Legal Considerations

Some regions may require permissions or licenses for keeping, breeding, or rearing swans. The proposed rearing site must comply with the local and national wildlife laws and regulations. It would be advisable to contact the local wildlife agency to understand the legal implications before setting up the rearing site.

4.2 Available Space and Natural Resources

Swan rearing comes with its own unique set of considerations and requirements in terms of available space and natural resources to ensure the health, comfort, and well-being of these elegant birds. Space and natural resources are essential aspects that can significantly impact the success of a swan rearing facility. In this long content, we will discuss essential factors that will help you plan and optimize the available space and natural resources for swan-rearing endeavours.

Available Space Considerations

Adequate space is pivotal for swans' well-being, enabling them to indulge in their natural behaviors while ensuring a harmonious and healthy living environment. Given below are some factors you need to assess when determining the necessary space for swan rearing.

Range of Movement

Swans need ample space to swim, feed, roam, and fly. The size of the available space should be large enough to accommodate a sufficient range of movement based on the following needs:

- For swimming, aim for a pond or lake with at least half an acre of surface area.

- Land surrounding the water body should provide enough area for the swans to roam, forage, and engage in preening and other behaviours.

Space Requirements for Multiple Pairs

Swans are known to become territorial during the breeding season. If you plan to introduce additional pairs or different waterfowl species, extra space is required to ensure each pair can establish their territory without conflicts. Ideally, each pair would need at least an acre or more to create their territory.

Natural Resources Considerations

Natural resources encompass water, food sources, nesting materials, and protection. Utilizing and preserving these resources will directly impact the success of swan-rearing activities.

Water Quality

Maintain a high-quality water source that ensures swans' health and supports the ecosystem. Water bodies should be clean, with minimal contamination from chemicals, pollution, or excessive algal growth. Regular water quality assessments are recommended to maintain a suitable environment for swans.

Natural Food Sources

Swans primarily rely on aquatic plants and invertebrates for nourishment. The water body and its surroundings should be rich in these natural food sources to support the swans' diets. Some natural food sources for swans include submerged plants, seeds, insects, small fish, and algae.

Supplemental Feeding

While the goal is for swans to rely on natural food resources, there might be a need for supplemental feeding in specific cases such as harsh weather or an inadequate supply of aquatic vegetation. This can include poultry layer pellets, grains, corn, or greens, such as lettuce or spinach.

Nesting Materials

Natural nesting materials available in the surrounding habitat should be sufficient for the construction of nests. The following types of materials are typically favored:

- Aquatic plants (reeds and sedges)
- Twigs and small branches
- Grasses

Protective Measures

Catering to the swans' protective needs entails offering natural and artificial resources to ensure their safety from predators and harsh weather conditions, such as:

- Natural barriers (dense vegetation, islands)

- Artificial structures (shelters, fencing, floating nesting platforms)

- Man-made roosting sites (trees or structures)

Sustainable Management of Natural Resources

Lastly, it is crucial to adopt sustainable management practices that help conserve the available natural resources while ensuring the health and well-being of the swans. Some of these practices include:

- Regular habitat monitoring (water quality, vegetation, and predator control)

- Implementing pest control measures (insects, invasive vegetation)

- Proper waste disposal (regular cleaning of waste materials)

- Conservation-oriented development (habitat extensions, restoration of wetlands)

4.3 Providing Food and Water

Swans are iconic waterfowl known for their majestic beauty and elegant behaviour. Keeping and caring for swans in a managed setting requires an understanding of their dietary needs and preferences. These birds have unique dietary habits that significantly contribute to their health, their feather condition, and their overall well-being. This long content will explore the various aspects of swan diet, including their feeding behaviors, natural diet, supplemental feeding, and other special dietary considerations.

Natural Feeding Behaviors

Swans are highly adaptable and versatile in their eating methods. They exhibit a variety of strategies to obtain their food, which primarily involves feeding in water but can also include some foraging on land. Swans feed by:

- Dabbling: This involves upending in shallow water to reach plants growing underwater while their body floats on the surface.
- Grazing: Swans are known to graze and forage on land, particularly consuming grasses and agricultural crops.
- Diving: Some individuals might dive to access deeper aquatic vegetation, although this is less common.

Natural Diet of Swans

Swans are essentially herbivores with the bulk of their diet consisting of aquatic and terrestrial vegetation. Here are some key elements of their natural diet:

- Aquatic plants: Swans largely feed on a variety of submerged and floating aquatic plants, including pondweed, waterweed, stonewort, and algae. They consume both the leafy parts and the roots.

- Terrestrial plants: When feeding on land, swans consume a variety of grasses, grains, and legumes. They are also known to eat agricultural crops such as wheat and corn.
- Invertebrates: While plants form the majority of a swan's diet, they also consume some aquatic invertebrates, including insects and small crustaceans, which provide essential proteins.

Supplemental Feeding

In a managed setting, swans may require supplemental feeding, especially during adverse weather conditions or if the natural food sources become scarce. This supplementary diet can contain:

- Corn: Cracked or whole kernel corn is an excellent energy source for swans, it is easily digestible and well-liked.

- Lettuce and Greens: Fresh lettuce, spinach, and other leafy greens offer beneficial vitamins and fiber.

- Poultry Pellets: These are specially manufactured complete pellet foods designed for waterfowl that provide balanced nutrition.

- Fruit and Vegetables: Swans can also eat a variety of fresh fruits and vegetables which provide additional vitamins and antioxidants.

Spotlight: Bread – A No-Go: Despite popular belief, bread should not be given to swans. It offers little nutritional value and can lead to health problems such as angel wing disease.

Dietary Special Considerations

Along with understanding what to feed your swans, it's crucial to recognize specific dietary requirements and restrictions to ensure their optimal health:

- Grit: Swans need grit (small rocks and pebbles) to help grind food in their gizzard, improving digestion. This can be naturally available in their habitat or provided if needed.

- Water for Drinking: Clean, fresh water should always be available for drinking. Swans need to take water into their beak when eating dry food to help them swallow.

- Feeding Juveniles: Cygnets (baby swans) require high protein in their diet for healthy growth. Specialized waterfowl starter feed can be used.

Dietary Needs and Preferences of Swans

Swans are majestic waterfowl that require specific dietary considerations to maintain their health and well-being in both natural and captive environments. Understanding their dietary needs and preferences is crucial for providing them with a balanced and appropriate diet. In this comprehensive guide, we will explore the natural diet of swans, their feeding behaviours, supplemental feeding, and special dietary considerations.

Natural Diet of Swans

Swans are herbivorous birds that mainly consume a variety of aquatic plants and grasses, which provide the necessary nutrients for their growth and survival. Here are some essential elements of their natural diet:

- Aquatic Plants: Swans primarily feed on submerged and floating aquatic plants such as pondweed, waterweed, and stonewort. They consume the leaves, stems, and roots of these plants, deriving important nutrients from them.
- Terrestrial Plants: In addition to aquatic vegetation, swans also graze on various terrestrial plants, including grasses and agricultural crops. They are

known to feed in fields and meadows, consuming grains, grasses, and legumes.

Feeding Behaviors of Swans

Swans employ different feeding strategies depending on their environment and available food sources. These behaviours include:

- Dabbling: Swans dabble by extending their necks underwater and upending their bodies to reach submerged vegetation. This behavior allows them to feed on aquatic plants growing in shallow waters.

- Grazing: Swans are capable of grazing on land, primarily consuming grasses and other plant species. They may forage in pastures, meadows, and agricultural fields, utilizing their long necks to reach the vegetation.

- Filter Feeding: Some swans, particularly the smaller species, are filter feeders. They feed on small invertebrates, algae, and other microscopic organisms by straining water through specialized lamellae in their beaks.

Supplemental Feeding

In instances when natural food sources are scarce or during extreme weather conditions, supplemental feeding may be necessary to ensure the swans receive adequate nutrition. Some common supplemental feed options include:

- Poultry Layer Pellets: These commercially available pellets are formulated to provide a balanced diet for waterfowl, including swans. They consist of a mixture of grains, seeds, and other essential nutrients.

- Leafy Greens: Fresh lettuce, spinach, and other leafy greens can be offered as a supplemental food source. These provide additional vitamins and minerals to support the swans' overall health.

- Fruits and Vegetables: Swans can also benefit from a variety of fresh fruits and vegetables, including apples, strawberries, carrots, and peas. These can be chopped into small pieces or grated for easy consumption.

Special Dietary Considerations

There are several specific dietary considerations when caring for swans:

- Grit: Swans require small gravel or grit in their diet to aid in digestion. Grit helps break down food in their gizzard and facilitates the absorption of nutrients. Provide access to a source of grit, such as small stones or crushed oyster shells.

- Water: Swans need access to clean, fresh water for drinking and foraging. They often take water into their beaks while feeding on dry food to aid in swallowing and digestion.

- Protein Requirements: Juvenile swans, known as cygnets, have higher protein requirements for proper growth and development. It is essential to ensure they receive adequate protein through their diet, either from natural sources or supplemented feed.

Chapter five

Swan Behavior and Communication

5.1 Social Behavior within Swan Flocks

Swans are known for their graceful appearance and regal demeanour. However, their beauty extends beyond their physical attributes; they also exhibit complex social behaviours that contribute to their overall appeal. Whether in their natural habitats or in captivity, swans exhibit a wide range of social behaviours that are worth exploring. In this guide, we will delve into the intricacies of social behaviour within swan flocks.

Social Structures within Swan Flocks

Swans are social creatures that typically exist in large flocks. These flocks are highly structured and organized, with individuals taking on various roles. Typically, a pair of swans forms the nucleus of a flock, with other individuals joining them to form a larger group. Swans are known for their monogamous mating systems, with pairs remaining together for life.

Within the flock, there is typically a hierarchy based on age, size, and perceived strength. Older and larger birds typically take the lead, dictating the movements of the flock and protecting the younger, more vulnerable members of the group.

Communication and Vocalizations

Swans use a variety of vocalizations to communicate with each other. They may hiss, grunt, or trumpet to convey various emotions and messages to their flock mates. Some common vocalizations include:

- Trumpeting: Swans trumpeting is typically associated with aggression towards other birds or animals. They may trumpet when they feel threatened or to warn other birds of danger.
- Hissing: Hissing is a more gentle vocalization used to communicate with other swans. They may hiss to greet each other, signal that they are busy foraging, or to establish territorial boundaries.
- Grunting: Grunting is another common vocalization used by swans. They may grunt while feeding, while resting, or during courtship displays.

In addition to vocalizations, swans also use body language to communicate. They may use open wings, lowered heads, or raised necks to signal their emotions and intentions to other birds.

Aggression and Conflict

While swans are typically peaceful and non-aggressive, they can become territorial and protective during certain times of the year. During breeding season, swans may become more aggressive towards other birds, particularly those that are perceived as a threat to their eggs or young. They may also become more territorial during nesting, chasing away other swans that come too close to their nest.

In the wild, conflicts can arise between different flocks over food, nesting spaces, or territory. These conflicts are typically resolved through displays of aggression, such as trumpeting or lunging at other birds. However, these conflicts are usually short-lived, and once a hierarchy is established, the flocks coexist peacefully.

Social Bonding and Family Structures

Swans are known for their strong, monogamous pair bonds that can last a lifetime. During breeding season, pairs work together to construct a nest, incubate eggs, and raise their young. Both parents participate in these activities, with males typically taking on the role of protector and defender.

5.2 Hierarchical Structure and Interactions

Swans, with their elegance and grace, also exhibit fascinating hierarchical structures and intricate social interactions within their flocks. These hierarchies play an essential role in their daily lives, influencing their behaviours, resource access, and overall flock dynamics. In this guide, we will delve into the hierarchical structure and interactions among swans, shedding light on their social complexities.

Hierarchy Formation

Swan flocks generally have a hierarchical structure, with dominant individuals occupying higher positions and exerting influence over others. Hierarchy formation often begins during the swans' early development, as they grow and interact with others within their flock. Factors such as age, size, and strength contribute to the establishment of these hierarchies. Older and larger swans tend to assume dominant positions, while younger or newly arrived individuals occupy lower ranks.

Dominance Displays and Interactions

Within swan flocks, dominance is usually displayed through various behaviours and interactions. Dominant swans assert their position through displays of aggression, vocalizations, and body language. These displays serve as a means of communication and a way to establish and maintain the hierarchy within the group.

Some common dominance displays and interactions among swans include:

- Neck Threats: Dominant swans may extend their necks forward and lower their heads, indicating their dominance and intention to maintain control. This display is often accompanied by vocalizations, such as hissing or grunting.

- Wing Threats: Swans may partially open their wings and hold them in a raised position as a form of threat display. This behaviour aims to intimidate and establish dominance over other individuals.

- Aggressive Postures: Dominant swans may exhibit aggressive postures such as upright positioning and swimming towards other flock members. These displays are meant to assert their dominance and discourage others from encroaching on their territory or resources.

- Chasing and Pecking: Dominant swans may engage in chasing or pecking behaviours to assert their dominance and maintain order within the flock. They may drive away subordinate individuals or prevent them from accessing food sources.

It's important to note that while dominance displays and interactions occur, swans typically maintain a peaceful coexistence within their flocks. Aggressive behaviours are usually limited to brief encounters and serve as a way to establish and reinforce the hierarchy rather than causing harm.

Benefits of Hierarchy

The hierarchical structure within swan flocks serves several important purposes:

1. Resource Allocation: The hierarchy helps regulate access to limited resources such as nesting areas, foraging grounds, and mates. Dominant swans

typically have priority access to these resources, ensuring their survival and that of their offspring.

2. Flock Coordination: The hierarchical structure enables efficient coordination and coordination within the flock. Dominant individuals often take the lead in determining the flock's movements, warning others of potential threats, and ensuring collective safety.

3. Social Stability: By establishing a clear hierarchy, swans minimize conflict and maintain social stability within the flock. This reduces the frequency and intensity of aggressive encounters and promotes a more harmonious coexistence.

Flexibility and Changes in Hierarchy

Swan hierarchies are not rigid and can change over time, particularly during certain phases of the swans' life cycles.

As younger swans mature and grow in size and strength, they may challenge the dominance of older individuals and strive for higher positions in the hierarchy. These challenges can lead to reshuffling of ranks and adjustments in the flock's social dynamics.

5.3 Body Language and Swan Gestures

Beyond their striking beauty and enchanting presence, swans communicate through a rich repertoire of body language and gestures. Their elegant movements and subtle signals convey important messages and play a vital role in their social interactions and overall behaviour. In this guide, we will explore the fascinating world of body language and gestures in swans, shedding light on their nonverbal communication techniques.

Head Movements and Positions

Swans can convey a range of emotions and intentions through their distinct head movements and positions:

1. Head Bobbing: Swans often bob their heads up and down rhythmically, particularly during courtship displays. This movement is accompanied by soft vocalizations and serves as a way to attract mates and establish their presence.

2. Head Tucking: When swans tuck their heads close to their bodies, it signifies relaxation and contentment. They may exhibit this behaviour while preening, resting, or simply enjoying a peaceful moment.

3. Head Raising: Swans raise their elongated necks in a graceful manner to showcase their size and dominance. This behavior is often seen during aggressive encounters or territorial disputes as a way to intimidate rivals or assert their authority.

Wing Movements and Displays

Swans also utilize their wings to communicate various messages and intentions:

1. Wing Flapping: Wing flapping is a common behaviour in swans, especially during takeoff and landing. However, they may also employ this gesture in other situations, such as to release excess energy, signal excitement, or warn off potential threats.

2. Wing Water Slapping: Swans vigorously slap their wings on the water's surface as a defensive gesture. This behaviour is exhibited when they feel threatened or when trying to protect their young or territory. The loud noise created by the wing slapping can act as a deterrent to potential predators or intruders.

3. Wing Fluttering: Wing fluttering is a subtle movement in which swans gently vibrate their wings without fully extending them. This behavior is often associated with courtship displays, where it adds to the swan's elegance and attractiveness.

Body Postures and Positions

Swans use their body postures and positions to convey various emotions and intentions:

1. Upright Position: A swan standing upright with its body tall and sleek exudes confidence and dominance. This posture is often seen during territorial displays or when asserting their authority within the flock.

2. Arched Neck: When a swan arches its neck gracefully, it signifies curiosity or alertness. They may adopt this posture when investigating their

surroundings or when something catches their attention.

3. Feathers Display: Swans can erect and fluff up their feathers, creating an impressive display known as "fanning." This behavior can communicate aggression or act as a defense mechanism to appear larger and more intimidating when faced with potential threats.

4. Sleeping Position: Swans sleep by folding one leg beneath their bodies while tucking their heads into their feathers. This position allows them to rest while maintaining some level of alertness and protecting themselves from potential dangers.

Eye Contact and Gaze

Eye contact and gaze also play a role in swan communication:

1. Direct Eye Contact: Swans often engage in direct eye contact with their fellow flock members, especially during interactions or encounters. This can convey a range of emotions and intentions, such as curiosity, friendliness, or aggression.

2. Avoiding Eye Contact: On the other hand, swans may avert their gaze or avoid eye contact as a sign of submission or respect towards dominant individuals. Such behaviour acknowledges their lower rank in the flock's hierarchy and helps maintain social order.

Chapter six

Conclusion

6.1 Recap of Key Points

Rearing swans is a rewarding yet challenging endeavour that requires care, patience, and attention to detail. These elegant and majestic birds demand specialized care and attention to ensure their health and well-being. In this guide, we have covered various aspects of swan rearing, from habitat management to feeding, health, and breeding. This recap will summarize the critical points covered in the guide to help you succeed in rearing swans.

Habitat Management

Swans require a clean, well-maintained habitat to thrive. Here are the key points to consider when creating and managing their habitat:

1. Provide a freshwater pond or lake with adequate space for swimming, nesting, and foraging.
2. Ensure proper water quality through regular testing and treatment.
3. Keep the habitat clean, free from debris and pollution.
4. Provide natural vegetation, such as reeds and grasses, as well as artificial nesting platforms and shelters.

Feeding and Nutrition

Swans require a balanced and varied diet to maintain their health and vitality. Here are the key points to consider when feeding swans:

1. Offer a combination of commercial swan feed and fresh, natural foods such as greens, vegetables, and aquatic plants.
2. Avoid feeding bread and other processed foods that lack proper nutrition and can harm swan health.

3. Divide feed into multiple small portions and feed at regular intervals throughout the day.

4. Ensure access to clean, fresh water for drinking and bathing.

Health and Care

Maintaining good health and hygiene is essential for swan well-being. Here are the key points to consider when caring for swans:

1. Regularly inspect swans for signs of illness, injury or abnormal behavior.

2. Provide proper shelter and protection against harsh weather conditions.

3. Practice good hygiene practices, such as keeping the habitat clean and sanitizing feeding areas and equipment.

4. Partner with a qualified veterinarian with avian experience to establish a health care plan.

Breeding

Breeding swans require specialized care and attention to ensure successful reproduction. Here are the key points to consider when breeding swans:

1. Ensure proper nutrition and habitat conditions for breeding pairs.

2. Provide a secure and protected nesting area.

3. Monitor nesting progress regularly and provide additional care and protection during incubation and hatching periods.

4. Seek professional guidance from experienced breeders or avian veterinarians.

6.2 Final Thoughts on Rearing Swans

Rearing swans is a fascinating and rewarding experience, but it requires dedicated effort to provide these magnificent birds with the care they need to thrive. The key to successful swan rearing lies in understanding their unique needs and behavior and implementing proper care and management practices to ensure their health and well-being. In this guide, we covered various aspects of swan rearing, including habitat management, feeding and nutrition, health and care, and breeding. This final thoughts section will provide additional insights and considerations to enrich your swan rearing experience.

Understanding Your Swans

Swans are complex birds that require attentive care and management. Understanding their behaviour and needs is critical to their health and well-bcing. Here are some additional considerations to keep in mind:

1. Get to know your swans on an individual basis. Each bird has a unique personality and behaviour patterns that can help you tailor your care and handling practices accordingly.

2. Observe your swans closely to detect subtle signs of illness or stress. Early detection can help prevent health issues from worsening.

3. Allow your swans to express their natural behaviour, such as foraging, preening, and swimming. Providing an enriched environment promotes physical and mental health.

Navigating Challenges

Rearing swans can present a range of challenges, from disease outbreaks to aggressive behaviour. Here are some strategies to manage potential issues:

1. Develop a contingency plan for managing potential disease outbreaks, such as quarantining affected birds, contacting a veterinarian, and following proper biosecurity measures.

2. Address aggressive behaviour appropriately, such as by separating conflicting birds or providing more space.

3. Stay informed about swan management best practices by participating in swan rearing communities, attending workshops and conferences, and seeking advice from experienced breeders.

Sharing Your Passion

Rearing swans can be a fulfilling experience, and many enthusiasts enjoy sharing their passion with others. Here are some ideas for doing so:

1. Host educational events for the community, such as swan watching tours or lectures on swan behaviour and conservation.

2. Volunteer with conservation organizations or wildlife rehabilitation centers to help care for injured or orphaned swans.

3. Share your experiences and information about swan rearing through social media, blogs, or other online platforms.

www.ingramcontent.com/pod-product-compliance
Lightning Source LLC
Chambersburg PA
CBHW062343290526
45794CB00005B/2096